WHERE IS THE STRAW?

Written by

Read With You Science Academy

Read With You

The characters and events portrayed in this book are fictitious. Any similarity to real persons, living or dead, is coincidental and not intended by the author.

Published by Read With You Publishing
Designed by Read With You Science Academy

ISBN-13: 978-1-64616-114-0
ISBN-10: 1-64616-114-9

Second Edition April 2021
Printed in the United States of America.

Welcome to the

READ WITH YOU SCIENCE ACADEMY!

Today we will do an experiment called

WHERE IS THE STRAW?

Before we begin, let's review some important words:

surface:
The upper layer of an area of land or water.

to disappear:
To stop being visible; cannot be seen.

narrow:
Long and not wide; small from one side to the other side.

QUESTION:

What I want to know.

Will the straw look like it's broken when placed in a liquid?

- ☐ Yes, the straw will look like it's broken when placed in a liquid.
- ☐ No, the straw will not look like it's broken when placed in a liquid.

HYPOTHESIS:

What I think will happen and why.

I think the straw **will** look like it's broken because...

I think the straw **will not** look like it's broken because...

The straw **will** look like it's broken because:	The straw **will not** look like it's broken because:
The liquid will cause the straw to look as if part or all of it is missing.	The liquid will not change the appearance of the straw at all.
The liquid will dissolve part of the straw.	It's impossible to break a straw.

What do you think?
Circle your hypothesis!

EXPERIMENT:

How we will test our hypothesis.

THINGS WE NEED

- A straw and a pencil
- A tall, clear drinking glass
- Cooking oil, such as vegetable oil or olive oil
- A clear bowl or pan (in case the water spills.)
- Paper towels or a towel (for clean up!)

straw

glass

cooking oil

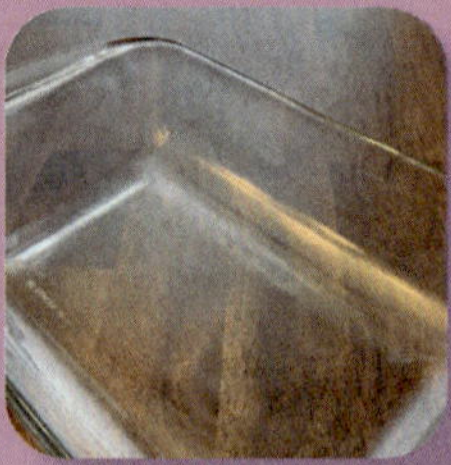

bowl

PROCEDURE:

The steps to complete the experiment.

Step 1A

- Pour $^{1}/_{3}$ cup of oil into the glass.
- Remember to put the glass over a bowl or in a pan in case your **liquid** spills!

Step 2A

- Place the straw inside the glass.
- At the **surface** of the oil, does the straw look like it has been cut into two pieces?

Step 3A

- Lean the straw against the side of the glass.
- If the straw does not lay flat against the side, **bend** the straw slightly at the bottom to help it stay in place.

Step 4A

- Slowly turn the glass and observe the straw. Does the straw appear to become wider or **narrower**? Can you make it **disappear**?
- Remember: Use a towel or paper towel to clean any liquid on the table or floor!

SAFETY FIRST

Write down what you see

In your science journal,

write your results!

- What was your hypothesis? (Look on page 5 of this book!)
- Was your hypothesis correct?
- What happened to the straw?

Let's do it again!

If we use a pencil in place of a straw, will it look the same? Let's do an experiment to find out. You will need the same materials as before, but this time we will use a pencil in place of a straw.

QUESTION:

What I want to know.

Will the pencil look like it's broken when placed in a liquid?

HYPOTHESIS:

What I think will happen and why.

I think the pencil will look like it's broken because...

I don't think the pencil will look like it's broken because...

What do you think?

Write your new hypothesis in your science journal.

PROCEDURE:

The steps to complete the experiment.

Step 1B

- Remove the straw from the glass containing oil.
- Remember to put the glass over a bowl or in a pan in case your liquid spills!

Step 2B

- Place the pencil inside the glass.
- At the surface of the oil, does the pencil look like it has been cut into two pieces?

Step 3B

- Lean the pencil against the side of the glass.

Step 4B

- Slowly turn the glass and **observe** the pencil. Does the pencil appear to get wider or narrower? Can you make it disappear?
- Remember: Use a towel or paper towel to clean any liquid on the table or floor!

Write down what you see

In your science journal,

write your results!

- What was your hypothesis? (Look on page 12 of this book!)
- Was your hypothesis correct?
- What happened to the pencil?

Describe/draw what the straw and pencil looked like from each view:

Outside:		Inside:		As I turned the glass:	
Straw	Pencil	Straw	Pencil	Straw	Pencil

Why did this work?

- This experiment worked because of **refraction**.
- Refraction is the bending of light as it moves from one material to another. This is what makes items inside a glass of water look different and strange.
- Oil bends light even more than water. It can bend light so much that a straw or pencil resting against the side of the glass can be hidden from your eyes if you hold the glass a certain way.
- Can you think of other examples of refraction?

SCIENCE

- When you looked through the glass at the straw, the straw did not appear to be straight. In fact, it looked as if it had been cut into two pieces and separated by the liquid. This is an **optical illusion** because the straw was not cut, nor did it disappear.
- Light can either **reflect** or **refract**.
- Refraction is the bending of light as it passes from one **transparent** substance into another. Light refracts whenever it travels at an angle into a substance with a different **refractive index**.
- This change in direction is caused by a change in speed. For example, when light travels from air into water, it slows down, causing it to continue to travel at a different angle or in a different direction.
- When the straw appeared to be broken or disappear, it was due to refraction. We see refraction commonly in water, and as we saw in our experiment, oil bends light even more than water.

GLOSSARY

To bend: To curve out of a straight line or position.

To disappear: To stop being visible; cannot be seen.

Narrow: Long and not wide; small from one side to the other side.

To observe: To watch something carefully.

Optical illusion: Something that looks different from what it really is.

To reflect: To move in one direction, hit a surface, and then quickly move in a different and usually opposite direction.

Refraction: The bending of light rays as they move from one material into another material.

Refractive index: The ration of the speed of radiation (such as light) in one medium (such as air, glass, or a vacuum) to that in another medium.

Surface: The upper layer of an area of land or water.

Transparent: A substance that light can pass through without scattering.

NEW IDEAS

How could we change this experiment?

- Could we mix oil and water in the same glass and perform the experiment?
- Could we use a glass of clear liquid soap and perform the experiment?

Can you think of any ideas to improve the experiment?

- Use different types of liquids and compare the straw and pencil.
- Use different quantities of liquids and compare the straw and pencil.

JOURNAL

Made in United States
Orlando, FL
07 November 2026

90337527R00017